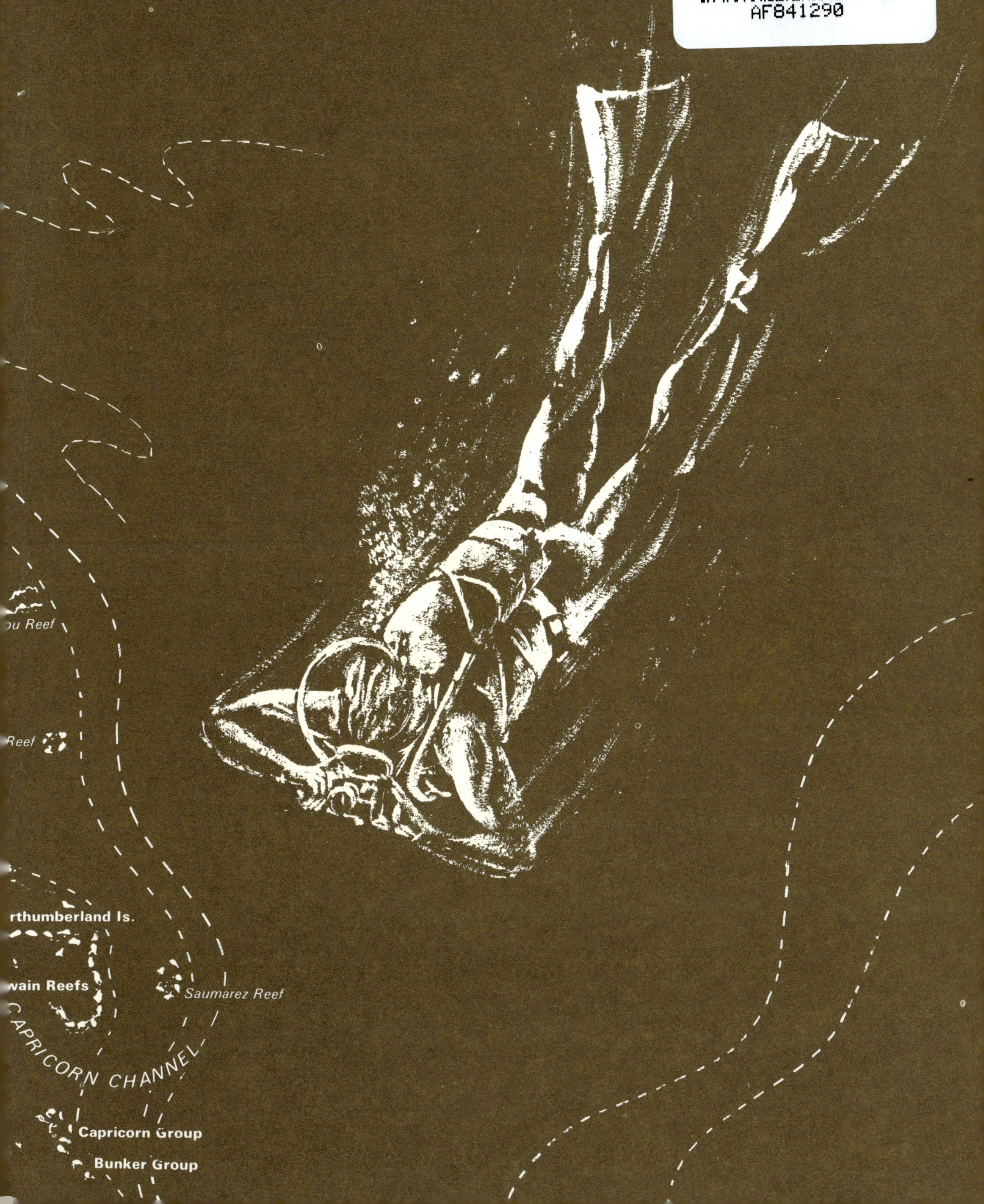
ou Reef
Reef
rthumberland Is.
wain Reefs
Saumarez Reef
CAPRICORN CHANNEL
Capricorn Group
Bunker Group

Natural Life of the BARRIER REEF

Walter and Jean Deas

This book is dedicated to the memory of Don Morser, Jim McKirdy, and Dave Powdrill: three friends. Their tragic deaths on the Reef, 22 January 1972, will never be forgotten.

RIGBY

RIGBY LIMITED • ADELAIDE • SYDNEY • MELBOURNE • BRISBANE • PERTH

Library of Congress Catalog Card Number 70–160563
National Library of Australia Registry Card Number & ISBN 0 85179 442 4
Wholly designed and set up in Australia

Printed in Hong Kong

Natural Life of the Barrier Reef

Off the coast of north-eastern Australia is an area of 80,000 square miles of sea, about 100 feet deep and studded with coral reefs. The Great Barrier Reef extends some 1,260 miles from the Tropic of Capricorn northwards along the Queensland coast, almost reaching to the shores of New Guinea. A zoological and scenic wonder, it contains the densest assemblage of living organisms to be found in any region of comparable size in the world.

The reef is not a continuous coral wall, but is an intricate labyrinth of reefs of various shapes and sizes, split by deep and often treacherous channels. In the more northern parts from Thursday Island to Cairns the reefs are close together and quite near to the mainland. South of this area the reef becomes less definite, a mass of irregular scattered reefs and cays which diverge from the coast line until they are nearly two hundred miles from the coast. Here along the outer edge, the reef drops away to nearly 1,000 fathoms.

This distance from the mainland explains Lieutenant James Cook's failure to discover the reef until he had sailed many miles to the north. On 11 June 1770, the reef became a part of Australia's history when his ship the *Endeavour* went aground on the reef which now bears the vessel's name.

As far as is known the next white man to see and sail through the reef was Captain Bligh, who was cast adrift by the *Bounty* mutineers in 1789. His party was exhausted and near starvation when they went ashore on an island which he named Restoration Island, before continuing on the epic voyage to Timor.

The reefs and islands of these perilous waters have undoubtedly had earlier visitors. The Chinese claim to have discovered the reef more than 2,000 years ago. The Japanese give accounts of sea voyages to the reef in the fifteenth century. However, there is little evidence to substantiate these reports.

Captain Luis de Torres, the Spanish navigator, sailed the treacherous waters at the reef's extreme northern tip in 1605. Perhaps earlier navigators from the Old World penetrated these dangerous seas.

If one could believe the old Torres Strait pearlers and divers, Barrier Reef identities, and native legends, then many ancient ships have crashed on the coral fangs of the reef. Today it is difficult to ascertain what actually occurred in those perilous days of sail. But the anchors, cannon and other less perishable artifacts are mute testimony to this graveyard of the past.

Romantic, mysterious, rich in beauty and magnificent in proportions, the reef stands as a massive undersea monument to nature. This tract of sea is an underwater laboratory so immense that no one has ever completely documented its contents.

Two kinds of islands occur in the reef. Near the mainland are the high islands, the tops of mountains of a former range of coastal hills now submerged. The other islands are the coral sand cays which have been built up from the debris of the reefs and are surrounded by coral banks many times their area.

Tiny coral polyps building upwards to the sunlight have industriously formed the reefs. The seas have pulverised the coral into a solid mass above the waterline. The relentless force of sea and wind washes broken coral, sand and silt into the lee side of reefs where the accumulated debris forms a cay. Seeds borne by birds, wind and sea sprout in the decaying coral and sand. Grass and other small plants appear. Nesting birds fertilise the soil with their guano and trees and shrubs add to the vegetation. The result is a lovely green island contrasting strongly with the surrounding reef.

One of the reef's greatest attractions for the non-diving tourist is the reef at low tide. With a guide, one can walk out over the exposed coral. A myriad of shapes and colours are found in a complex of pools, terraces and crevices. For reef walking, stout soled shoes are required as coral scratches become infected quickly in the tropics.

One must peer into the coral pools to fully appreciate the abounding life. There are hosts of small striped and spotted fish, sea urchins, starfish, clumps of sponges, sea anemones, bêche-de-mer, and shells. The diversity of marine life will keep one enthralled until the tide starts to return.

But to see the true reef one must go down into the warm, pellucid water and enter a new world of wonder—a paletteful of fishes, terraces, colonnades, arches and domes of coral, and white clean sand. A skin-diver's mask, fins and snorkel are all that are required to enjoy the magic land beneath the surface. An advanced scuba diver can stay longer and deeper but the shallows alone can be a stunning visual experience. Coral, colourful tube worms, hydroids, all are a thousand models in an underwater studio.

Against this fantastic background of coral galleries and towers, countless varieties of fish of every shape, pattern and colour swim freely back and forth. Marauding schools of kingfish swoop by, chasing living clouds of pilchards. Approach a kaleidoscope of reef fish in a clump of acropora coral, and they sink slowly into its protection as the great finned monster hovers menacingly above.

Gliding by towering coral archways and around gigantic domes, one wonders what will next meet the eyes. A large nudibranch gliding over a coral plate, a massive coral trout being picked over by tiny cleaner fish, a coral head with nothing but multi-coloured tube worms. Then a swirl of water indicates that a shark is nearby, but with a quick movement it disappears. The sandy bottom moves as a stingray settles its bulk.

Angelfish, maori cod, butterflyfish, damselfish, turrum, and humbugs abound, and in these waters many more fishes can be seen in one dive than in a year in southern waters. A milling school of terrified small-fry hurtle past and head for the surface as trevally swoop in for a feed.

You notice a track on the sandy bottom. Hold your breath and swim down for a closer look. A small highly polished olive shell creeping along has cut this furrow. From the surface you can now suddenly see tracks criss-crossing everywhere, showing that there are other shells around.

A group of huge niggerheads of living corals tower upwards from the sea bed. They look like huge underwater castles. Here reef fishes parade in their incredible finery. They come in multiple shades of yellow, red, blue and green streaked and striped as though a crazed artist had been let loose underwater.

Many fish can be tamed and in certain locations on the reef an underwater guide will show you how to hand feed the fish. Some can be aggressive, especially the maori cod. The damselfish are pugnacious, the butterflyfish are timid, but a blue angelfish, although normally extremely shy, will come up and inspect your hand, look into your mask with the most expressive face, and one feels guilty if no food has been taken along to feed it. The noise from thrashing fins and scuba exhaust bubbles, even the pressure waves from a diver's moving body, will scare fish. Generally it is a rare sight to see a fish eating from a diver's hand.

As you paddle around this populous reef, you may find the water filled by the massive bulk of a twelve-foot manta ray. Tiny cleaner wrasse swarm all over it removing parasites. Fishes of all sizes and patterns poise and extend their fins as they are picked over by these small cleaner wrasse.

The most stunning underwater scenes are found in the Coral Sea Cays. Willis Group, Marion Reef, Saumarez Reef, Kenn Reef, and others are all 200–300 miles off the coast, and are seldom viewed by tourists. On the surface, birds are the only inhabitants of the tiny saucer-shaped islets. When disturbed they wheel and screech at the human intruders.

Below, a diver hangs suspended in a pool of living blue. The clarity of the water is dazzling. So far these secluded reefs have been saved from pollution. Here the coral is not so fragile nor do the fish seem so brightly coloured, but the unlimited visibility compensates for this. Shells abound, the most common being huge helmet and spider shells. Sea snakes and sharks of varied species are prevalent in this part of the reef. Fortunately the reef is virtually untouched except for a section of Saumarez Reef where the wreck of the 10,000-ton American liberty ship, the *Francis Preston Blair*, has been used as a target for practice bombing. Nearer the mainland, Michaelas Cay and Fairfax Island have also been used as bomb targets causing untold havoc to the fauna and ecology.

There are many threats to the reef—illegal fishing, pollution, vandalism, mining for limestone, and oil drilling. And apart from man's threats, the reef faces an endless battle for survival against the forces of nature and predators, such as the Crown of Thorns Starfish, *Ancanthaster planci*, which is in plague proportions on some southern parts of the reef.

Insufficient scientific work has been carried out on the reef and it is only now that the reef's potential is being realised. For the tourist trade alone it is a great economic asset worth millions of dollars each year.

Conservationists are deeply concerned about the widespread despoliation of the reef. We must not let man ruin this legacy. The Great Barrier Reef belongs not just to Australia but to the world, and it must be safeguarded from destruction.

No descriptive account can adequately carry to the mind the grandeur of this vast coral rampart and the life allied with it. It was a realisation of this which motivated the authors to obtain the photographs for this book.

Walter and Jean Deas

Below: Jean and Walter Deas

Top left: The Tournefortia-tree, *Messerschmidia argentea,* is a low-lying tree which colonises parts of the coral cays

Bottom left: The Pandanus Palm, *P. tectorius,* offers cool shade from the hot sun. The fruit grows in pineapple-like clusters, and is capable of floating long distances. This is why the trees often grow at the water's edge

Above: The Pisonia tree, *Pisonia grandis,* is a dominant species on many islands and cays. The sticky seeds adhere to the White-capped Noddy Terns which disperse them far and wide in their travels

Left: White-capped Noddy Terns, *Anous minutus,* nesting in the Pisonia trees. They are extremely numerous in the Capricorn and Bunker Groups of islands

Facing page
Top left: A Mutton-bird or Wedge-tailed Shearwater, *Puffinus pacificus,* near its underground burrow at night. A summer breeder on many islands throughout the reef

Top right: At a depth of ninety feet two divers set out to explore their spectacular surroundings

Facing page
Bottom left: A Masked Gannet, *Sula dactylatra,* broods two eggs. The nest is just a depression in the coral debris

Bottom right: A colourful sponge from deep water. Sponges represent one of the oldest forms of animal life

Top: Shallow water sponges encrusting on the reef. In calm water many sponges grow into a branched form

Bottom: Underwater movie photographer using 16 mm Beaulieu at Lamont Reef. On his back is a battery pack to power his movie light

Above: Like a turreted castle this tube sponge juts out from the reef. Sponges often harbour other animals, mostly crabs and worms, in their tube system

Above: A large clam shell sits on the sea bed with its mantle full exposed

Left: A sea snake at Kenn Reef. Sea snakes are generally mild-tempered reptiles, although there are exceptions. Usually they are inquisitive but after a quick inspection they will swim away. However, we have seen one chase a diver from a depth of eighty feet and attempt to bite him during the fifty-yard swim to the boat

Above: A sea snake at Saumarez Reef

Left: A colourful nudibranch, a member of the mollusc family. These strange creatures are slugs without shells. They have external gills of various shapes on their backs

Left: A tiny uncommon cowrie, *Cypraea asellus.* They are usually located under coral boulders but can be seen out in the open at night

Left: Jean brings two large Helmet Shells, *Cassis cornuta,* up from their sandy confines

Facing page: Jean with a small turtle which had been stranded. It was released as the incoming tide quickly approached

Right: A labyrinth of coral pools on the reef flat at Heron Island. A multitude of marine life swarms beneath the still waters

Below: Holiday-makers laze in the warmth of the tropic sun

Top right: The sea pounds on the edge of a coral rampart. Note the clearly defined surge channels

Bottom right: Inch-high tube worms, *Spirobranchus giganteus,* resemble twin Christmas trees. Their delicate branches are the gill plumes and can be found in an amazing variety of colours

Above: A large (30") sluggish bêche-de-mer crawls across the sandy sea floor. Some holothurians feed on small sea creatures, while others feed on sand and extract organic remains or small organisms. Some species of bêche-de-mer, or trepang, have a commercial value as food favoured by the Chinese and some islanders of the south-west Pacific

Facing page: Feather stars, survivors of a group of animals abundant in ancient fossils, lie amid a variety of sponges

Right: A large Crown of Thorns Starfish, *Acanthaster planci*. These large echinoderms have become a plague on certain parts of the reef. They destroy large coral formations by feeding on the polyps

Above: Like a bunch of white feathers the stinging hydroid, *Lytocarpus philippious,* hangs from the cliff face

Top of facing page: The vivid blue starfish, *Linckia laevigata,* is a conspicuous reef dweller. This starfish seems to prefer the reef flats and is seldom seen in the deeper waters

Bottom of facing page: A common starfish, *Fromia elegans.* This small brightly coloured species is found in the shallows, often out in the open

Top left: Jean picks up the starfish, *Nardoa novae-caledoniae*. This would be the most common starfish found on the reef flat. It usually rests on the sandy bottom at the edge of coral pools

Top of facing page: Slate-Pencil Sea-urchin, *Hetero-centrotus mamillatus,* has heavy blunt spines and this species is difficult to locate amid the coral reefs

Bottom left: The formidable spines of the black spiny sea urchin, *Diadema setosum,* protect it from predators

Bottom of facing page: The red-eyed crab, *Eriphia sebana,* is a common, nocturnally active crab on the beach front amid the intertidal rocks

Top right: Gorgonia corals are usually found in deep water, often under ledges and in caves, though their resilient forms can also be located on open cliff faces. This is a fan of gorgonia with a crinoid attached to it

Bottom right: This sphere of coral has the convolutions of the human brain, so is naturally called Brain Coral

Left: The "stinging" coral, *Millepora,* takes on a number of forms. It can often be recognised by its texture and yellowish colours. It is not a true coral but a hydrozoan

Below: A small branching coral, *Stylaster* sp. Often located in caves and under overhangs

Above: The expanded polyps of the coral, *Tubastrea* sp., sometimes appear during the day

Top of facing page: The mushroom coral, *Fungia fungites,* is fairly common on the reef flat and over the reef edge

Bottom of facing page: A Butterfly Cod, *Pterois volitans.* These fish have poisonous dorsal spines which can inflict a very painful sting. Some species are extremely dangerous to handle

Left: The Pennant Coralfish, *Heniochus singularis*. This fish is often confused with the Moorish Idol, *Zanclus cornutus*, as both are very similar in shape and colours

Below: The Beaked Butterflyfish, *Chelmon rostratus*, is one of the common *Chaetodons* of the reefs

Top right: The Six-barred Angelfish, *Euxiphipops sexstriatus.* The brilliant colours and patterns of the angelfishes have captured the attention of diver and scientist alike. The change in colour patterns from the juvenile stage to adult has confused many people, and different names have been given to the same species

Bottom right: Black-backed Butterflyfish, *Chaetodon melanotos.* Fairly rare in some locations, this *Chaetodon* grows to about seven inches

Left: The electric ray is harmless unless touched or molested

Below: A Coral Cod, *Cephalopholis miniatus*. It is usually found in caves and under overhangs

Left: The Clown Trigger-fish, *Balistoides niger,* is perhaps the most elegant member of its family

Below: Coral Trout, *Plectropoma maculatum,* is a large reef predator which is much sought after as a food fish by fishermen

Above: Jean hand feeds a Blue Angelfish, *Pomacanthus semicircularis*. These beautiful fish usually live in pairs. The young tend to be very differently coloured

Top of facing page: A shark passes close to the photographer

Bottom of facing page: The sleek shape of the Bronze Whaler Shark often inspires enthralment rather than dread

Jean with an eight-foot Tiger Shark, *Galeocerdo cuvieri*. This is a dangerous shark which is easily recognised by its blunt head and characteristic tiger-stripe markings

PAPUA
Murray Is.
TORRES STRAIT
Thursday I.
Cape York
Cape Grenville
Cape Direction
Raine I.
CORAL SEA
100
500
1000
2000
Claremont Is.
Princess Charlotte Bay
Flinders Group
CAPE YORK PENINSULA
Cape Melville
Cape Flattery
COOKTOWN
Endeavour Reef
Port Douglas
Holmes Reef
Diane Rock
Willis Is.
Magdelaine Cays
Herald Cays
Trinity Bay
Michaelmas Cay
Green I.
CAIRNS
Flinders Reefs
Dunk I.
Hinchinbrook I.
Orpheus I.
Palm Is.
Magnetic I.
TOWNSVILLE
Whitsunday Group
Holbourne I.
Hayman I.
Bowen
Proserpine
Whitsunday I.
Lindeman I.
Brampton I.
Percy Is.
MACKAY
QUEENSLAND
ROCKHAMPTON